water falling from the sky,

water in the sea

Boats float on it,
Bricks sink through it,
Fish live under it,
And we all splash in it!

Almost all the fresh, unsalty water on Earth is either deep under ground or frozen as ice or snow.

Contents

97 per cent of the water on Earth is salty sea water.

Only one per cent
of the water on Earth
is available for us
to drink, but that
is still plenty for
all our needs.

That leaves only three per cent
of Earth's water that is fresh.

droplet

I know what water is!

Water

Everything that is alive is made of water, and that goes for carrots, camels, and crabs too.

Animals

Animals like you, me, and camels need to drink water to stay alive. We need to drink plenty every day.

We are
per cen

You would only live for three days without water.

Water is the most precious thing we have on Earth.

4

You **drink** water and you **wash** in water. But did you know that 75 per cent of you **is** water?

bubble

75 water

Even your bones are half water.

Plants

Carrots and other plants drink by sucking water up from the ground through their roots.

How much?

Make a note of how much water or other liquid you drink in a day. I bet it's quite a lot.

Everything that is living on Earth needs water to survive.

You and I have to hold our **breath** when we dive. But some animals, like fish, can breathe underwater using special flaps called gills.

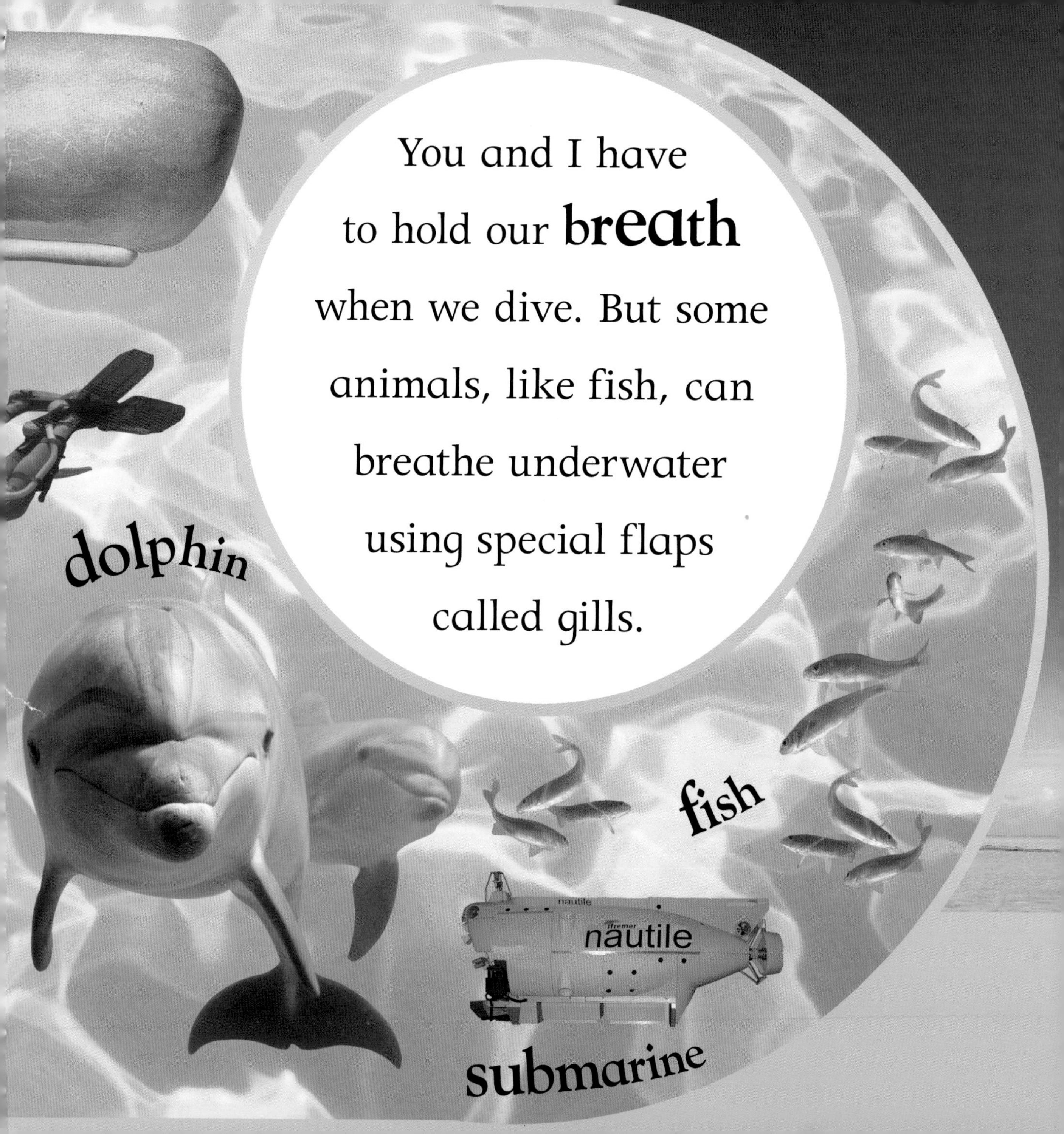

dolphin

fish

nautile

submarine

whale

divers

When divers go underwater,
they have to take big bottles
of air on their backs, or
travel in submarines.

eel

shark

seahorse

octopus

Whales have
to swim to the
surface to take
a breath.

A watery world

There is **more** water on the surface of the
Earth than there is land – 66 per cent is
covered in **seas** and **oceans**. We can't
live under water, but there are plenty
of animals and plants that can.

Nobody can drink sea water because it is
too salty. We can only drink fresh water.

We're so cool!

Fruit and yoghurt go hard when they **freeze** because they are partly made of water. Try freezing a banana – does that go hard too?

Ice-pops

Ice isn't just for winter, it's a great
way to cool down in the summer too.
Add ice cubes to your drink, or treat
your friends to these **cool** lollies.

Mix it up

Mash up your
favourite fruit
with orange juice
or runny yoghurt.

Pour it out

Push sticks into
lolly moulds, then
pour in the fruit
mixture.

Pop it out

Put them in the
freezer overnight,
then enjoy your
home-made lollies.

Ice

When water gets very cold, it **freezes** and turns into ice. Have you ever held an **ice cube**? It's wet and slippery and will begin to melt in your warm hands.

Who likes ice?

9

Some areas of the world are made of ice. The Arctic, near the north pole, is just a huge lump of ice floating in the ocean. The ice never melts at the Arctic.

Krill like to live in the cold, dark waters near the bottom of the sea.

Icicles

Look out for **icicles** on windowsills next winter. They are made when rain freezes while it drips.

Icebergs

Icebergs are huge chunks of solid snow that float in the cold seas near the north and south poles.

Emperor penguins live in Antartica, near the south pole.

The sun sucks up water from the ground into the air and that makes clouds.

rain hat

Did you know that you drink the same water that the dinosaurs drank millions of years ago?

dinosaur

Have you ever noticed that puddles vanish when the sun comes out?

wellington boots

Rainwater ends up in our taps. It's the water we drink.

puddles

Have you ever stood in fog?
Fog is just a very low cloud.

When water falls from the sky, it's called rain. But where does rain come from?

Clouds are tiny drops of water that hang in the sky. When the clouds cool down, the water falls as rain.

It's raining!

DANGER!
Slippery

hail

Sometimes hail falls in lumps the size of golf balls. You don't want to be standing under those.

When it is very cold, rainwater freezes in the air and falls as snow or hail.

Snow and ice are very slippery so watch out for those slippery slopes!

Sometimes when it rains, the sun is shining too. That's when you can see a **rainbow**. When sunlight hits the raindrop it turns it into a rainbow of colours.

Water journey

When it rains the water must go somewhere or we would be living in a flood. Where does it go?

In towns and cities, water flows down drains in the roads to pipes underground.

Most rivers travel over land...

When you flush your toilet or empty the bath, where does the water disappear to?

12

The sewers (huge underground pipes),
not only collect rainwater but also all
the water that goes down your plug holes.

The sewers deliver the water
to reservoirs to be recycled
into drinking water.

River flow

Most rainwater runs down into the ground or ends up in rivers that flow to the sea.

Reservoirs

Some river water and underground water ends up in huge lakes called reservoirs.

Turn the tap

Reservoir water is cleaned to make sure it's safe to drink, then it flows through pipes to your tap.

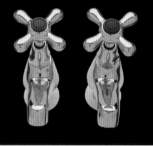

...until they meet the sea

Waste water from your home flows into huge pipes underground called sewers.

Turn on the tap

Are you feeling thirsty? Turn on the tap and pour yourself a **drink** of water. Now you know that this water has travelled hundreds of miles to reach you.

The average person uses enough water in one to day to fill over 400 drinks cans!

Too much,

Although rain is good, too much of it is dangerous. When a town **floods**, the rushing water can be strong enough to wash houses away.

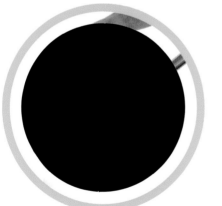

When a town or city floods, the water often mixes with dirt, so it isn't safe to drink or even use for cooking.

DANGER! FLOOD

What do we use water for?

watering

plants

Remember that all these activites use water. You'll find you use water far more than just for drinking. We couldn't live without it.

washing clothes

giving the garden a drink

brushing teeth

If you leave the tap running when you brush your teeth, you lose a bucketful.

washing-up dishes

bathing bodies

flushing

the toilet

cooking food

A washing machine uses 12 buckets of water each time it washes your clothes. That's a lot!

For one whole day, make a note of all the water you use in your home. And that includes **flushing** the toilet and brushing your teeth, as well as drinking.

15

too little

Imagine waking up and finding you have **no taps**. That's how lots of people live in the world. They have to fetch water in other ways.

Some people have to walk hundreds of miles to fetch water. They have no taps, no hosepipes, and no flushing toilets.

hose pipe

In some countries people only have one bucket of water to live on for a whole day.

bubbles

Air is lighter than water, so anything that is full of air will float!

swimming

The more you move your arms and legs in the water, the more you can stay afloat.

Bathtime!

Water is weird stuff. We have to learn to swim or we sink, yet huge boats float on it. Why is this?

Have you ever noticed that however many bubbles there are, they are always round?

18

What's inside the
hull of a boat?

Air!

boat

Even heavy
things will float
if they're flat enough
to cover a large
area of water.

19

Rubber ducks are full of air and very light so they bob on top of your bath.

...or swim?

Remove the clay, press it flat, and curve the sides up so it looks like a little boat. Now place it on the water. Does it float?

Flat wins!

Your boat floats because the flat shape spreads its weight over the water's surface. And what's inside? AIR!

Air makes things float

Soap bubbles are full of air, which is why they sit on top of water.

A ping-pong ball is full of air, so it floats. A boat is also full of air, which helps it float.

A golf ball is heavy, with no air inside, so it sinks.

Bath play

Roll a solid lump of modelling clay into a round ball and drop it into a bowl, a sink, or even a bath full of water.

Sink...

Because the ball is solid clay, with no room for air, it sinks straight to the bottom.

LONDON, NEW YORK, MUNICH,
MELBOURNE, and DELHI

Written and edited by
Penelope Arlon
Designed by
Tory Gordon-Harris

DTP Designer: Almudena Díaz
Production: Claire Pearson
Publishing managers:
Sue Leonard and Jo Connor

First published in Great Britain in 2006 by
Dorling Kindersley Limited
80 Strand, London, WC2R 0RL

A Penguin Company

2 4 6 8 10 9 7 5 3 1

Copyright © 2006 Dorling Kindersley Ltd

A CIP catalogue record for this book is available from
the British Library.

ISBN-10: 1-4053-1312-9
ISBN-13: 978-1-4053-1312-4

Colour reproduction by Media Development and
Printing, United Kingdom
Printed and bound in China by
Hung Hing Printing Co., Ltd

Discover more at
www.dk.com

Index